ARSON PROSECUTION:

ISSUES AND STRATEGIES

FEDERAL EMERGENCY MANAGEMENT AGENCY

UNITEDSTATES FIRE ADMINISTRATION

This publication was supported by the Federal Emergency Management Agency/United States Fire Administration grant No. EMW-86-C-2080. Its contents do not necessarily represent the policy of the funding agency. Moreover, the opinions and views expressed herein should not be construed to represent the policy of Abt Associates or the International Association of Fire Chiefs.

TABLE OF CONTENTS

TABLE OF CONTENTS
(continued)

INTRODUCTION

This document is intended to provide concise practical and technical guidance on arson prosecution. It draws heavily on state arson prosecutors' guides and other literature, and on presentations made at the first in a series of conferences for arson prosecutors at the Federal Law Enforcement Training Center in Clynco, Georgia in November 1985 which was jointly sponsored by the National Institute of Justice (NIJ), U.S. Department of Justice and the Bureau of Alcohol, Tobacco and Firearms (ATF), U.S. Department of the Treasury. These presentations have been separately summarized in an NIJ report (Hammett, 1987).

This document also contains a guide to key literature on arson prosecution. This guide, found in Appendix A, suggests specific readings for prosecutors desiring additional information on the topics discussed in this document. Full citations to materials cited in the text and in the guide to sources by topic may be found in the references at the end of the report.

ARSON INVESTIGATION AND PROSECUTION CAN BE SUCCESSFUL

Many prosecutors believe that arson cases are heavily circumstantial and very hard to win without specialized legal and technical expertise. Many arson cases do rely on circumstantial evidence and involve technical issues of fire behavior. However, research shows that prosecutorial success is eminently achievable in arson cases. On the other hand, prosecutors may encounter particular difficulties in arson cases. Jurors may have difficulty understanding the cases and judges may even be hostile to them, believing that, absent death or injury, such cases belong in civil court with the insurance company or property owner as plaintiff. As a result of these perceptions, it may be difficult to get arson cases accepted for prosecution. Inexperienced prosecutors may be particularly reluctant to accept arson cases,

Eyewitnesses to arson are rare and arson motives are often difficult to identify and substantiate. Moreover, the incendiary origin of the fire is often difficult to prove. Arson is unique in that the prosecutor must not only prove that a crime was committed (by establishing incendiary origin) but also disprove other possible explanations (by eliminating all accidental causes) (Maryland State's Attorneys Coordinator, 1981).

At first glance, the statistics of arson investigation and prosecution suggest that arson is a difficult crime to investigate and prosecute. A National Institute of

1

Justice study of four cities found that suspects were identified in less than one-third of the cases studied and in only four percent was any adult convicted of a crime. (Hammett, 1984). Many of the unsolved fires occurred in vacant buildings with no witnesses and no other information available to investigators. Such cases may be essentially unsolvable.

Uniform Crime Reporting figures from the FBI show that only 19 percent of reported arsons were cleared by arrest in 1985. This is comparable to the clearance rates for property crimes such as burglary, larceny-theft, and motor vehicle theft, but much lower than rates for other violent crimes such as murder, rape, and aggravated assault.

On the other hand, research in some cities shows that once an arson case is accepted for prosecution, the conviction rate is at least as high as for other felonies. While many of the cases accepted for prosecution in the NIJ study were simple and straightforward, many others that resulted in prosecution and conviction were complex, circumstantial cases. More than one-third of all convictions in this study were based entirely on circumstantial evidence -- that is, cases with neither eyewitness testimony nor a confession (Hammett, 1984).

There are many examples of entirely circumstantial arson cases that went to trial and resulted in convictions, The following brief case summary suggests the kinds of evidence that can bring a conviction in the hands of a skilled prosecutor:

> On the basis of three key pieces of circumstantial evidence, the owner of a house was convicted of burning it down: 1) through his own admission, he was placed at the scene shortly before the fire; he claimed he was washing dishes and someone threw a firebomb through the window; 2) the prosecutor showed that no curved pieces of glass, as would come from a broken bottle, were found in the room--only flat pieces of glass; and 3) most important, pieces of ceiling tile burned on the top but not on the bottom and a nail used to secure the tile in the ceiling not burned on the head but with soot along its shaft, together suggested that the ceiling had been pulled down prior to the fire to facilitate its spread. This conflicted with the defendant's contention that the ceiling was intact at the time of the fire. (Contributed by James Larkin, Assistant District Attorney, Suffolk County [Boston], Massachusetts.)

THE KEY ISSUES OF ARSON PROSECUTION

Convictions can be obtained in arson cases if prosecutors and investigators work cooperatively, and pursue cases with imagination and aggressiveness. The key issues for arson investigators and prosecutors are how to develop and maintain such cooperation, imagination, and aggressiveness. Some of the specific questions to be explored in this document are the following:

- when and how should the prosecutor become involved in the arson investigation?--e.g., should prosecutors attend fire scenes and, if so, which fire scenes?

- what are the legal and constitutional requirements of a proper fire scene search?

- how can the prosecutor help ensure that a careful and thorough scene examination is conducted?

- how much detailed knowledge of fire cause-and-origin determination must prosecutors possess to do their jobs properly?

- what are the best ways to document the fire scene?

- what are the key types of witnesses to identify and interview?

- what are the best types of motive evidence and how can they be developed?

- what is the optimum degree of specialization in arson prosecution?

- what are appropriate screening criteria for arson cases?

- what is the most effective mode of charging--preliminary hearing or grand jury?

- what statutes are applicable to arson case charging?

- what are the most effective ways of preparing witnesses for trial?

- what types of jurors are likely to be most responsive to the government's arson case?

- how can the prosecutor make best use of the opening statement and closing argument?

- What is the best sequence of prosecution witnesses?

- What interrogation tactics are most effective for particular witness types?

- How can the prosecutor most effectively meet the most common defenses in arson cases--e.g. alternative theory of cause-and-origin, alibis, intoxication, mental incapacity, insanity?

- What procedures apply to juvenile arson cases?

On many of these issues, there appears to be broad consensus among experts, while on others significant disagreement exists. This document presents the key points of agreement and summarizes the key arguments where there is disagreement.

ROLE OF THE PROSECUTOR IN ARSON INVESTIGATION

Perhaps the most important ingredient in successful arson prosecution is effective cooperation and coordination between arson investigators and prosecutors. Each brings a unique and critically important perspective and set of technical skills to bear on a case. The investigator understands fire and arson investigation and the prosecutor understands the legal requirements of arson prosecution. Failure to take account of these perspectives or to mesh these skills can spell doom for an arson case.

Too often, however, this cooperation and coordination have been missing. The NIJ study found that, in most jurisdictions, prosecutors are rarely involved in arson cases until their formal presentation, In other words, the prosecutor normally does not even see the case until it is fully developed and presented for screening (Hammett, 1984). At the same time, researchers have discovered rather low rates of investigative success in arson cases. The NIJ study discovered that suspects were identified in only 29 percent of a four-city sample of arson fires (Hammett, 1984), while another found that nearly twenty percent of a sample of cases drawn in eight cities were aborted due to errors or oversights by investigators IMcClees, 1983). Both studies found that arrests generally occurred early in an investigation, if they occurred at all. The one city of those studied in which prosecutors were regularly involved in arson cases prior to arrest exhibited high arrest and conviction rates compared to the other cities in the sample. By contrast, the city in which arson investigators had the least contact with prosecutors prior to case screening also exhibited the lowest conviction rates (McClees, 1983). These findings point up the need for early coordination of effort between investigators and prosecutors.

But what is the proper scope of such coordination? Should prosecutors attempt to take control and supervise investigations from the start? This would likely dissatisfy investigators. Alternatively, should prosecutors limit their involvement to consultation and rendering of legal advice upon request?

State arson prosecution guides generally agree that the prosecutor should be involved early in the case, ideally by attending the fire scene (e.g. California District Attorneys Association, 1980; Burnette and Smith, 1980). Attendance at fire scenes and other forms of early involvement in arson cases help the prosecutor in the following ways (California District Attorneys Association, 1980 and Toscano in Hammett, 1987):

- allows the prosecutor to see the evidence when it is fresh;

- prepares the prosecutor for effective description of the fire to the jury;

- provides information for effective direct and cross-examination of cause-and-origin witnesses;

- helps the investigators at the scene "think like prosecutors" by exposing them to the prosecutor's evidentiary needs and concerns;

- makes the prosecutor available to answer investigators' legal questions regarding

  --search and seizure

  --interrogation of witnesses

  --arrest procedures, and

  --evidence chain of custody;

- helps ensure that all necessary evidence is collected and preserved;

- helps ensure that all appropriate witnesses are identified and interviewed;

- helps ensure all relevant reports are obtained;

- helps to maintain the continuity of case development;

- ensures that cases are screened for weaknesses at an early point;

- ensures that every effort is made to remedy identified evidentiary problems before the case reaches trial;

- helps ensure that proper priorities are established among competing cases; and

- helps ensure that all suspects and leads receive timely followup action.

Clearly, prosecutors cannot, and probably need not, attend all fire scenes. It is particularly important for the prosecutor to be involved from the very beginning in complex arson-for-profit cases--to begin building such cases for trial starting at the fire scene and to evaluate evidence in the field with a constant view to its ultimate presentation in court (Connecticut Fire Inspector Joseph Toscano in Hammett, 1987). Some objective criteria for prosecutor call-out should be established based on the following:

- fire seriousness (death, injury, number of alarms, estimated dollar loss);

- early assessment of case complexity (e.g. possibility of fraud or profit motive, complex scene examination issues);

- presence of challenging legal issues (search and seizure, security of the fire scene);

- identification of important witnesses; or

- possibility of a confession.

Moreover, jurisdictions should define the scope of prosecutors' involvement in investigations. Based on organizational and experiential factors, jurisdictions should try to specify the prosecutor's role at the fire scene. Should they:

- direct the investigation?

- participate in the investigation? or

- merely act as legal advisors on evidentiary and search-and-seizure issues?

Examining the Fire Scene

The success or failure of an arson case may depend on the quality of the fire scene examination. As legal advisor to the investigator, the prosecutor can and should play an important role in this process. According to North Carolina state arson investigator David Campbell, "if you don't have a good cause-and-origin determination, you don't have corpus delecti, and if you don't have corpus delecti you don't have anything" (Campbell in Hammett, 1987). If a quick confession is obtained, investigators may neglect to conduct a thorough scene examination. This is extremely risky, however, since the confession will be useless absent proof that a crime was committed.

Many arson cases have failed because the prosecutor could not convince the jury that the fire was an arson. An effective scene examination requires that investigators work cooperatively with prosecutors and firefighters to secure the scene and conduct the search according to current legal requirements, and carefully develop evidence of the fire's cause and origin.

Legal Issues Regarding Fire Scene Search

When securing the fire scene and conducting their search, fire investigators must understand and act in accordance with current caselaw. The two leading cases in this area are Michigan v. Tyler [436 U.S. 499 (1978)] and Michigan v. Clifford [464 U.S.

287 [198], The major points in these cases are summarized in Figure 1. These cases clarified some legal points regarding fire scene search but, as Figure 1 shows, left other key points still unclear. Prosecutors must stay abreast of current caselaw in this area.

Several state arson prosecution guides contain good discussions of the <u>Tyler</u> case and guidelines for conducting fire scene searches in accordance with the <u>Tyler</u> doctrine (Texas District and County Attorneys Association, 1980; and Burnette and Smith, 1980). The following 'rules of thumb" are drawn from the Texas guide and from the recommendations of experienced arson prosecutors (Texas District and County Attorneys Association, 1980; ATF Regional Counsel Martin in Hammett, 1987):

- Investigators should get to the scene as quickly as possible and stay as long as possible.

- Continuous security of the scene should be maintained so there can be no later suggestion of tampering with evidence.

- Investigators should conduct as thorough a search as possible as quickly as possible so as not to overrun the "reasonable" period permitted under caselaw.* If the investigators must leave the scene before the examination is completed, the same personnel should return to continue the investigation.

- Investigators may wish to start the search in the area of the building considered least likely to have been the origin of the fire and finish with the areas considered most likely to have been the origin. Thus, it may be possible to search virtually the entire premises before encountering evidence requiring a warrant.

- If the investigators must leave the scene before the cause-and-origin determination is completed, they should obtain an administrative warrant before they return, rather than rely on the uncertain "continuing investigation" theory. If they do attempt to return without a warrant, the same personnel should return.

- If investigators have to leave after discovering evidence of arson, they should obtain a criminal evidentiary search warrant before returning. Due to the uncertainty regarding the definition of probable cause, investigators should assume that it exists when the nature of the investigation changes from general fire investigation to collection of arson evidence.

- One way to avoid legal problems in a fire scene search is to obtain a clearly drawn owner's consent. It is important that the consent document be very clear as to who is permitted to conduct the search and what period of time is covered. Finally, in order for the consent to be proper, all owners with a right to privacy in the

---

*For a discussion of caselaw on this point, see 4 ALR 4th 194.

Figure 1: The Tyler and Clifford Decisions

	Tyler	Clifford
During Fire suppression	No warrant required	Same as Tyler
"Reasonable time" after extinguishment of fire[a]	Investigators may remain to search for evidence of fire cause	Same as Tyler
	Investigators may seize evidence in "plain view" (such evidence will be admissible)	Same as Tyler
After expiration of "reasonable time"	Warrant[b] or written consent of owner required to continue search	Same as Tyler
"Continuing" Investigation	Definition unclear -- first warrantless return was proper but subsequent returns were improper absent a warrant	If investigators ever leave, they cannot return without a warrant
Discovery of evidence constituting "probable cause" of arson[c]	Not applicable	When such evidence is discovered, warrant or consent is required

---

[a] Definition of "reasonable time" is unclear. For a discussion of caselaw on this point, see 4 ALR 4th 194.

[b] Type of warrant required is unclear, Generally speaking, an administrative warrant should be obtained in order to continue a cause-and-origin investigation and a criminal warrant should be obtained for additional investigation after evidence of arson has been discovered at the scene.

[c] Definition of such evidence is unclear.

building must be identified and must sign the document. Connecticut fire investigator Joseph Toscano notes that he has never been refused when he asked an owner's consent (Toscano in Hammett, 1987).

- The safest approach is to obtain a warrant at the outset.

Prosecutors' offices and investigative agencies should work together to develop policies for fire scene security and evidence collection that take account of the key caselaw in this area. This is an evolving area which requires regular attention if potentially damaging investigative errors are to be avoided.

One final legal issue regarding scene examination should be mentioned. Prosecutors must avoid becoming witnesses at the scene, and thus disqualifying themselves from prosecuting cases (Gelband, 1980). This situation may be avoided if the prosecutor is always accompanied by an investigator at the scene, in which case the investigator can always testify to any necessary fact or point of evidence.

Cause-and-Origin Determination

Prosecutors should be familiar with the key technical aspects of fire cause-and-origin determination. This enables them to:

- properly evaluate the case;

- offer guidance to the investigators on strengthening the evidence of incendiary origin; and

- present this evidence clearly and understandably in court.

Incredibly, there have been instances in which an arsonist confessed but the case was dismissed because a cause-and-origin determination had never been done on the fire.* An informed and involved prosecutor would probably have prevented this from happening.

The following summary of the essentials of cause-and-origin determination is intended to familiarize prosecutors with key aspects of cause-and-origin determination. (North Carolina fire investigator David Campbell in Hammett, 1987).

- Investigators should **"read the scene" from the outside in**--that is, they should begin with examination of windows and doors for signs of forcible entry and the exterior of the building for evidence of unusual fire load or burn patterns and then move to the interior to

---

*State v. Brown, 308 NC 181, 301 S.E. 2d 89 (North Carolina, 1983).

search for the fire's point(s) of origin and ignition factor.

- Investigators should **concentrate their search on the floor level which is most protected from fire damage** and thus most likely to yield evidence of the presence of accelerants.

- Investigators must conduct their cause-and-origin determination in light of the **basic principles of fire behavior,** i.e.,

    --the need for heat, fuel and oxygen, and

    --the tendency of fire to burn up and out, following the path of least resistance.

- Investigators should watch for the following **types of unnatural burning** which strongly suggest that an accelerant was used: (Campbell in Hammett, 1987):

    --**deep charring.** Standard depth is one inch for every 45 minutes of freeburning. (This guideline refers to white pine or similar wood; it should be adjusted for harder or softer woods.)

    --**"alligatoring"** (char patterns) on wood. The size of the blister suggests intensity of the heat, Alligatoring low with large welts is an unnatural burn pattern.

    --**"v-patterns"** up and out on the wall or other surface above the point of origin. A wide sweeping v-pattern indicates a slower burn based on available combustibles while a narrow or inverted V indicates a much faster burn based on an accelerant.

    --**holes in the floor,** especially with edges bevelled out (indicating the wicking effect of a flammable liquid), are almost always suspicious, considering fire's natural tendency to burn up and out.

    --**the condition of glass,** which can show where the fire was most intense. Different types of marking or breaking are associated with different temperature ranges and the presence of different types of flammable materials. For example, a light soot present on a window may represent hydrocarbons from a petroleum product.

- investigators should also watch for **other important indicators of incendiary origin,** such as:

    --**multiple points of origin** (although investigators should be sure to consider the possibility of a natural "flashover" in the course of a fire),

    --evidence of **"trailers"** or pours of flammable liquid, and

--**ignition devices.**

As noted earlier, arson cases are unique in that "the prosecutor bears not only the burden of proof but also the burden of disproof." That is, the prosecution must establish, within a reasonable doubt, not only that the fire was incendiary but also that it was not accidental. This means eliminating all accidental causes to the satisfaction of the jury (Maryland State's Attorneys Coordinator, 1981). This is one of the areas of the prosecution's case that is most often vulnerable to defense attack, and the prosecutor should go over this in detail with the investigator at an early point in case development. In their scene examination, investigators must pay particular attention to the following possible elements of accidental cause:

- flammable liquids which may be legitimately present;

- the configuration and condition of electric service and appliances; and

- the possibility that careless disposal of smoking materials was involved.

Documenting the Fire Scene: Photographs and Videotape

Investigators should carefully document their observations at the fire scene with still photographs and, if the scene examination is complex, with videotape. Photographs and videotapes should:

- be able to tell the story of the fire from beginning to end; and

- be carefully shot to present the clearest view possible of the factors leading to the investigator's conclusion regarding the cause and origin of the fire.

Photographs may also help to clarify investigative terms and techniques for the jurors (Burnette and Smith, 1980). A prosecutor at the fire scene may be able to recommend particular views and camera angles that will present the evidence to best advantage in court. However, investigators and prosecutors should be aware that, in some jurisdictions, telephoto and wide angle shots may be excluded unless their use can be specifically justified. Prosecutors may be able to provide legal guidance on the sufficiency of the justification for employing such photographic shots.

There is some disagreement on the relative value of color versus black-and-white photographs of the fire scene. Some investigators and prosecutors believe that color photographs are most effective, while others believe that black-and-white

10

provides better contrast and shows burn and fire patterns to best advantage. One experienced prosecutor believes that both color and black-and-white photographs should be taken of fire victims. Thus, if the judge excludes the more dramatic color photographs as prejudicial, the prosecutor will still have the black-and-white pictures as a backup (New Haven Assistant State's Attorney Mary Calvin in Hammett, 1987).

Collecting and Analyzing Samples of Fire Debris

Laboratory analysis of fire debris may yield powerful evidence of incendiary origin. However, there is some question as to how frequently samples are even collected or analyzed in arson investigations. A recent study of arson prosecution discovered a "surprisingly" low reliance on laboratory analysis of fire debris in establishing incendiary origin. Only about 7 percent of the cases studied included such evidence. Samples should be collected and analyzed in as many cases as possible, in order to:

- buttress the prosecution case; and

- demonstrate that a complete and professional fire scene examination was conducted (Hammett, 1984).

Investigators should employ proper techniques to collect and preserve samples of fire debris. Much effort may be wasted at the fire scene if investigators do not know where to look for the best samples. Experienced arson investigators believe that the most promising areas may be the following (Campbell in Hammett, 1987):

- low places such as beneath carpets, carpet pads and baseboards, and

- since flammable liquids are lighter than water, samples should be taken from the surface of standing water in a basement using a gauze pad or similar absorbent.

Samples must be packaged and preserved so that none of the liquid or vapor is lost or adulterated either before laboratory analysis or between analysis and presentation of the sample to the jury in court. For example, the ATF laboratory prefers the following (ATF Laboratory Director Daniel Garner in Hammett, 1987).

- clean paint cans,

- glass jars (mason jars), or

- polyester nylon bags.

Investigators must insure that the proper chain of custody is maintained and documented for samples of fire debris and all other evidence. All too often, important evidence is excluded because the prosecution cannot document chain of custody. Early involvement by the prosecutor can be helpful in ensuring that the chain of custody is preserved and properly documented.

The laboratory analysis of fire debris is a highly technical scientific field. While prosecutors probably need not possess detailed knowledge of all aspects of fire debris analysis, they should know enough of the essentials to be knowledgeable in examining and cross-examining witnesses on the subject. (For a concise discussion of the subject, see Gaensslen and Lee, 1986.) The director of the ATF laboratory estimates that, in about 85-90 percent of arson cases, gas chromatography is sufficient to identify the substance in a sample. However, in the remaining 10-15 percent of the cases, computer-assisted gas chromatography/mass spectrometry is necessary because the presence of pyrolysis products (products created by the process of burning) in the sample render the chromatogram uninterpretable. While the ATF laboratory generally can make a fairly specific identification of the substance (i.e. gasoline, light petroleum distillate, medium petroleum distillate, heavy petroleum distillate), some other laboratories may not be able to be so specific. However, investigators should inform laboratory staff of their views or suspicions concerning the identity of the substance, since analyses can be set up to seek out specific substances (Garner in Hammett, 1987).

Developing the Criminal Case

At the fire scene and during the criminal investigation, the prosecutor can assist the investigator in building a case which will be effective in the courtroom. It is important to emphasize at the outset that the criminal investigation and the cause-and-origin determination are not separate stages. Rather, the chances of investigative and prosecutorial success will be maximized if they proceed simultaneously and, indeed, interdependently. The scene examination may yield important clues for the criminal investigation and vice versa. For example, a particular type of fire may suggest or eliminate a particular motive or class of suspects (Pisani, 1982). Building the criminal case involves:

- identifying and interrogating witnesses;

- developing evidence of arson motive;

12

- preparing complete and well-organized investigative reports; and

- timing arrests to best advantage.

Identifying and Interviewing Witnesses

Key witness types include:

- firefighters; (especially "first-in" firefighters)

- eyewitnesses;

- property owners;

- spectators at the fire;

- victims;

- neighbors; and

- business associates.

It is desirable to interview as many witnesses as possible at the fire scene. There are a number of reasons for this:

- some witnesses may be reluctant to talk later;

- the information is fresher; and

- suspects have less opportunity to plan possible alibis.

Witnesses at the fire scene should be asked about any suspicious people or suspicious behavior observed at or near the scene immediately before or during the fire.

"First-in" firefighters should always be interviewed. They can often provide valuable information on the appearance of the fire and the condition of doors and windows.

Alibi witnesses should be interviewed as quickly as possible after they are identified. These interviews should probe the account in detail, thus providing as many opportunities as possible to check the internal consistency of these witnesses' statements. * Finally, investigators should record their interviews with suspects and

---

*The Bureau of Alcohol, Tobacco, and Firearms has developed standard questions for various witness types at fire scenes, including firefighters, property owners, insurance personnel, neighbors, and general observers.

potentially hostile witnesses. Recordings are often useful in catching and documenting inconsistencies in a witness' story (Galvin and Toscano in Hammett, 1987).

Prosecutors should encourage investigators to keep track of the subsequent whereabouts of all witnesses. This will facilitate recontacts for additional questioning as well as notification and followup regarding court appearances.

Developing Evidence of Motive

Motive is not normally a required element of the crime in arson cases. However, motive evidence has very important persuasive value in many prosecutions. This importance is heightened in arson cases because of the frequent lack of direct evidence linking the defendant to the crime (Texas District and County Attorneys Association, 1980).

A wide range of evidence types have been held admissible by courts in the establishment of arson motive. In **commercial arsons,** evidence of motive may include:

- depleted inventories;

- overvalued stock;

- poor business conditions;

- change in character of the neighborhood; or

- overinsurance.

In **residential arsons,** motive may be shown by:

- physical condition of the property (deterioration or other indications of disinvestment);

- problems in selling the building;

- owner's desire to move from the area; or

- evidence of domestic problems.

The Florida arson prosecutors' guide presents a useful summary of caselaw on the admissibility of these types of motive evidence (Burnette and Smith, 1980, pp. 4B-9 to 4B-11).

Insurance Evidence. Evidence of arson motive may be developed from many different sources. Perhaps the most promising source is the insurance industry.

14

Insurance companies may have a significant advantage over public investigators in the examination of the fire scene and the collection of information and documents from the insured because of the terms of the insurance contract. The contract explicitly grants the company the right to enter the premises to investigate fire cause and to examine the insured under oath (Attorney David Strawbridge in Hammett, 1987). Moreover, insurance companies may often have more resources than public agencies to apply to an arson investigation once they are convinced that it is worth pursuing.

Several potentially valuable types of materials may be available from insurance companies:

- policies (including all declarations and endorsements);

- applications;

- risk inspection reports;

- photographs of the property;

- records of premium payments;

- claims history;

- statements under oath; sworn proofs of loss;

- underwriting files;

- loss prevention reports; and

- investigative reports (Bruce Bogart, American Insurance Association, in Hammett, 1987).

Public officials are likely to be most successful in obtaining these documents from insurance companies if they work through the company's legal department and make all requests for materials in writing (Strawbridge in Hammett, 1987).Furthermore, all fifty states have now passed some form of Arson Reporting-Immunity Law. These laws are intended to encourage information exchange between insurers and public investigators by freeing companies of liability associated with releasing information on their insureds. Immunity laws have facilitated many arson investigations. However, many companies and public investigative agencies continue to be very cautious about sharing information (for a full discussion of immunity laws, see Hammett, [1987a]).

Insurance information may be useful not only in establishing motive, but also in uncovering inconsistencies useful to the state's case in any aspect of the suspect's account. Careful examination may often detect contradictions between information in insurance documents and the defendant's interview statements. Moreover, any false statements found in insurance applications (e.g., regarding previous fires, valuation of property) may also be useful to the prosecutor (Galvin in Hammett, 1987).

Other Evidence Sources. There are a number of other useful sources of motive evidence for the development of arson cases. The following types of documents may be particularly valuable: (Panneton in Hammett, 1987).

- bank records: mortgage information, loan applications, payment records;

- property conveyances: ownership history, recent ownership transactions, sales prices;

- property tax records: property valuation, arrearages, abatements, tax liens, tax foreclosures;

- income tax returns: under-reporting of income;

- incorporation records: information on corporations, partnerships, annual reports versus sales tax reports;

- court records: litigation, bankruptcy, divorce;

- code inspection reports: violations, citations, photographs and plans, building permits;

- fire department records: fire history, safety code violations.

- licensing records: liquor license applications, names of corporate officials/partners; and

- utility records: payment history, service interruptions.

Developing Investigative Reports

The value of an arson investigative effort--no matter how well it was conducted--may be seriously reduced if it is not effectively documented in reports presented to the prosecutor. In the preparation of their reports, investigators should try to think like prosecutors and consider the most effective way to present the evidence in court. By laying out the case logically and clearly in their reports, investigators will maximize the likelihood that it will be accepted for prosecution (Hammett, 1984). Arson investigation units should consider providing training on the preparation of effective case reports. Prosecutors should certainly play a key role in such training, explaining to investigators how their reports can be made most useful in the case preparation process.

Prosecutors may also ensure that all critical investigative tasks are assigned and accomplished, so that all evidence, exhibits and witnesses are fully prepared for trial.

Timing of Arrests

There is a particular need for coordination between prosecutors and investigators in the timing of arrests, especially in complex arson-for-profit cases. It may be risky to rush the arrest in an arson-for-profit case. Instead, the prosecutor should probably wait at least until the property owner files an insurance claim, so that, if he or she inflates the loss, it becomes possible to charge insurance fraud in addition to, or instead of, arson.

Moreover, it may be important to "turn" the torch into a state's witness and "wire" him so that incriminating statements made by the target of the investigation may be captured on tape. It takes time and patience to develop a case in this fashion and a premature arrest may irreparably harm a carefully and painstakingly developed investigation. (Bronx Assistant District Attorney Barry Kluger in Hammett, 1987). Another potential advantage of delaying arrest is that it allows repeated interviewing of suspects, thus increasing the probability that they will make false or inconsistent statements (Calvin in Hammett, 1987).

ARSON PROSECUTION

As stated at the outset, convictions can be obtained in arson cases despite the peculiar difficulties often associated with them. For example, a recent study of arson prosecution found that once an arson case was accepted for prosecution, the conviction rate was 79 percent--at least as high as for other felonies--although the conviction rate in cases reaching trial was only 58 percent, somewhat below equivalent rates for other felony case categories in most jurisdictions (Hammett, 1984). Another study discovered that 61 percent of the charged defendants were convicted (McClees, 1983). The first study attributed its high conviction rates to the fact that most cases accepted for prosecution were simple and straightforward. Complex fraud cases accounted for only nine percent of the prosecuted cases in the study.

This section discusses the major issues in arson prosecution, including:

- degree of specialization;

- prosecutorial screening;

17

- charging modes;

- preparation of witnesses; and

- trial tactics.

Arson Prosecution Structure

The major issue regarding the organization of arson prosecution is the degree of specialization necessary for effective prosecution of cases. If it can be accommodated within the overall structure of the office, a certain degree of specialization is probably desirable (Hammett, 1984). The California arson prosecutors guide argues that every office, no matter how small, should assign and train at least one attorney to handle arson cases. Moreover, the guide recommends vertical prosecution of arson cases--i.e., the same attorney or attorneys prosecute the case from start to finish (California District Attorneys Association, 1980). There are a number of arguments in favor of specialization (Hammett, 1984):

- the complexity and technical challenges presented by some arson cases;

- the greater likelihood of specialized prosecutors developing close and regular working relationships with arson investigators;

- the desirability of improving relatively low trial conviction rates;

- the increasing skill and experience of the defense bar in arson cases;

- the opportunity for specialized prosecutors to become familiar with the arguments and strategies used by defense cause-and-origin experts; and

- the opportunity for prosecutors to develop and maintain close working relationships with insurance companies.

Hammett (1984) identifies the following basic typology of arson prosecution structures:

- **Non-Specialized/Horizontal,** in which there is no specialization **by** case type and a different attorney handles each major stage of the case. This is the most economical structure for criminal prosecution but it affords no opportunity for the development of the specialized expertise or ongoing relationships with investigators which may be very important in complex arson cases,

- **Specialized/Vertical,** in which the same prosecutor or unit in the office handles all arson cases and each case is handled by a single attorney from start to finish. This organization allows for the maximum specialization, but it may in fact constitute "overkill" since many arson cases are simple and straightforward and really do not require a specialized prosecutor.

- **Non-Specialized/Vertical,** in which a single prosecutor handles each case from start to finish but there is no specialization by case type. The major weakness of this system is that complex arson-for-profit cases may receive no specialized handling and there is ostensibly little opportunity for expertise to develop. However, informal arson specialization may develop under this approach.

- **"Hybrid" prosecution,** in which a designated attorney or unit screens all arson cases, vertically prosecutes all complex cases and other cases presenting challenging legal or technical issues, and passes the rest on to the office's general crimes processing stream. Although this approach may only be feasible in fairly large offices, it seems to offer the best of both worlds--specialization and efficiency (Hammett, 1984). The hybrid structure also affords the greatest potential for developing consistent and realistic screening of arson cases and for building close productive relationships between prosecutors and investigators.

Prosecutorial Case Screening

The major issues regarding prosecutorial screening of arson cases are:

- the criteria for case acceptance;

- the timing of screening; and

- the structure of screening,

According to two recent studies, most arson cases presented to prosecutors were filed in court. One study found that 66 percent of all arson arrestees in the sampled cases were charged while, in another, 76 percent of adult arson cases presented for prosecution were accepted (McClees, 1983; Hammett, 1984). Thus, it might appear that there is little to fault in the screening process and that the most severe problems arise in the investigative phase--that is, developing a case to the point that it is worthy of presentation to the prosecutor.

However, upon further consideration, it appears that there may be some deficiencies in the screening of arson cases. Although prosecutors may disagree on the correctness of this approach, they generally take a very conservative approach to screening arson cases (Hammett, 1984). Fully sixty percent of filed cases in this study included direct evidence (in the form of a confession or eyewitness testimony) and the vast majority included circumstantial evidence in all three key categories: incendiary origin, motive, and opportunity. Rejected cases tended to .include circumstantial evidence but no direct evidence. However, one city in the study (Cleveland) represents an instructive exception. Here, a much higher percentage of filed cases were based entirely on circumstantial evidence, yet this city exhibited the highest conviction rate in the study. Based on this evidence, the study urges prosecutors to adopt a more venturesome approach in accepting and trying "marginal" cases which rest on circumstantial evidence (Hammett, 1984).

In addition, the adoption of formal and specific screening criteria for arson cases can help to reduce subjectivity and inconsistency in the screening process. Such criteria should constitute a systematic method of

- evaluating evidence in key areas such as incendiary origin, motive, and opportunity;

- assessing the reliability and credibility of witnesses and the evidence corroborating co-conspirator testimony; and

- taking account of categorical factors such as the seriousness of the fire (Hammett, 1984).

Some experienced arson prosecutors echo the call for more venturesome screening, while others urge a more cautious approach. The authors of the California arson prosecutors guide recommend that a case should be declined outright only if the commission of a crime cannot be established or the evidence is irreparably weak (California District Attorneys Association, 1980). Some prosecutors' offices accept a case if there is a "reasonable probability of conviction," meaning credible evidence of incendiary origin and one of the following: documented lies by the defendant, evidence of motive, or evidence of opportunity (Galvin in Hammett, 1987). However, in some jurisdictions, such an evidence combination is not considered to produce a reasonable likelihood of conviction. Obviously, prosecutors must always act in accordance with the standards applicable in their jurisdictions.

A common argument for venturesome screening is that it might help to deter arsonists--especially arsonists-for-profit. Many prosecutors believe that they have a responsibility to the public to heighten this deterrence by accepting--and winning--difficult cases (Calvin in Hammett, 1987). Moreover, this will produce more guilty pleas.

Interestingly, it is sometimes argued that a circumstantial case may be stronger than a case with direct evidence from an informant or a "torch" turned state's evidence because there is no problem with the evidence being tainted by ulterior motives. In other words, facts do not lie the way people can lie. (Kluger in Hammett, 1987).

On the other hand, prosecutors must be careful not to jeopardize their credibility with judges by prosecuting cases that are too weak, More importantly, they should never lose sight of their responsibility to protect potential defendants from unreasonable prosecution. In short, ventursome screening should never become reckless screening, and should always adhere to legal and prosecutorial standards of ethics.

It seems particularly useful to have prosecutors informally evaluate cases prior to their formal presentation so that investigators may take appropriate action to correct deficiencies. Based on a thorough review of potentially applicable offenses and their elements, the prosecutor may be able to suggest untapped avenues of investigation and additional statutes to explore in charging the defendant (California District Attorneys Association, 1980).

Finally, jurisdictions should develop a structure for arson case screening that best meets their organizational requirements while addressing the technical issues posed by arson cases. Recent research shows that development of close working relationships between investigators and prosecutors may be facilitated, and more effective and consistent case screening, by employing a centralized/specialized or at least a centralized system--that is, in which all cases are screened by the same attorney or unit in the prosecutor's office (Hammett, 1984).

Determining the Mode of Charging: Grand Jury Versus Preliminary Hearing

Statute and case law differ widely regarding modes of charging, Some states are essentially preliminary hearing/information states, some are essentially grand jury/indictment states, while others require both a preliminary hearing and a grand jury indictment, and still others leave substantial discretion in this matter to the prosecutor,

Most experts urge prosecutors to avoid the preliminary hearing in arson cases, if at all possible, and to take their case directly to a grand jury. The grand jury has the following advantages:

- the **secrecy** of the grand jury proceeding helps to protect the prosecution's case from exposure or damage. This is particularly important in delicate cases resting on the testimony of "turned" co-conspirators,

- the **investigative** powers of the grand jury are particularly well suited to the development of complex arson-for-profit cases requiring numerous witnesses and substantial documentary evidence (California District Attorneys Association, 1980).

On the other hand, if a prosecutor must go to a preliminary hearing, the following potential advantages should not be overlooked. The preliminary hearing may:

- screen out weak cases that should not go further;

- encourage better early evaluation of cases;

- induce pleas;

- test prosecution and defense witnesses;

- discover defense theories of cases; and

- preserve evidence; for example, if an eyewitness testifies and is cross-examined at the preliminary hearing and then fails to show up at the trial, his or her preliminary hearing testimony will probably be admissible.

It appears that one school of thought favors aggressive use of the preliminary hearing to test witnesses. On the other hand, some experts argue that only the minimum necessary to satisfy the burden of proof should be presented at a preliminary hearing so as not to provide the defense with early discovery. This is an important tactical choice for the prosecutor to make (California District Attorneys Association, 1980).

Statutes Applicable to Arson Prosecutions

Whatever mode of charging is dictated or selected, the prosecutor must be thoroughly familiar with all applicable statutes. The California arson prosecutors guide notes that charging philosophy differs from place to place, but generally urges charging the most serious and provable offense, together with other offenses the proof of which will tend to strengthen the case on the lead charge (California District Attorneys Association, 1980). Obviously, the range of available and applicable laws differs from jurisdiction to jurisdiction. This section discusses the types of laws that tend to be available in most jurisdictions (Hammett, 1987):

State Arson Statutes

Previous studies have strongly recommended that arson laws be toughened, that loopholes be removed, and that the profit be taken out of arson" by statutory means. These legislative changes have largely been accomplished. It remains for them to be aggressively implemented by investigators and prosecutors. Hammett notes that arson statutes in the four jurisdictions studied had been considerably tightened in recent years, particularly with respect to arson-for-profit. Another study of eight cities concludes that "[e]xisting loopholes neither drastically interfere with arson investigations nor regularly undermine prosecution. Far more important is the quality and quantity of the investigations that do take place, and of the prosecutions that follow." (Hammett, 1984; McClees, 1983)

The state arson prosecution guides cited in this report all contain god discussions of their states' arson statutes and the required elements of proof. Summaries of caselaw are also provided in these state guides.

Other Laws

Other laws besides the basic arson statutes may be useful in prosecuting arson. These include the following (Hammett, 1987):

- **Insurance fraud statutes:** These may be included in basic arson statutes or be separately enacted. They generally cover making false statements on applications, filing false claims, and the like.

- **Mail fraud statutes:** Federal mail fraud statutes may be applicable if any false statement or fraudulent document, such as an insurance application, proof of loss or claim was sent through the mail. Even if the document was hand-delivered or sent by a private delivery service, mail fraud may apply if a confirmation copy was returned to the sender by regular mail.

- **Reckless endangerment statutes:** In New York and some other states, reckless endangerment laws may be used if an arson charge is not possible for lack of evidence of intent. In addition, courts in some states have applied the reckless endangerment law to the situation of firefighters, thus circumventing the "firemens rule." This held that no one could be held liable for death or injury to a firefighter unless it was the result of an intentional act.

- **Racketeer-Influenced Corrupt Organizations (RICO)** statutes: State and federal RICO laws may apply to arson-for-profit conspiracies. The civil (treble damages) provisions, although often difficult and complex to apply, may offer a promising strategy for taking the profit out of arson and having a more lasting impact on the arsonist (Maryland State's Attorney Coordinator, 1981).

23

- Criminal and civil forfeiture provisions: Under the federal Comprehensive Crime Control Act of 1984, the assets of certain arsonists may be forfeitable.

Depending on the legal environment in their states, prosecutors may be able to employ other types of laws in arson cases,

Preparation of Witnesses

There is strong agreement among prosecutors on the need for thorough pre-trial preparation of all witnesses in arson cases. This should occur individually as well as collectively. Individual preparation can

- iron out weaknesses in the testimony of weaknesses; and

- prepare witnesses for particular lines of cross-examination.

It may also be helpful to gather all witnesses together in one room to explain how all their testimony fits together. Such group sessions may also bring out new information.

Careful preparation is particularly important for witnesses whose testimony will be technical or unusually complex, such as investigators testifying on the cause and origin of the fire and laboratory chemists testifying on the analysis of fire debris. Thorough pretrial preparation should:

- ensure that the prosecutor knows what the witness is and is not prepared to say on the stand; and

- prevent any unfortunate surprises at trial.

During preparation, each witness should be led through his or her testimony, ideally proceeding from point to point by means of photographs and other exhibits. This evidence will help the witness keep the overall theory of the case, as well as key details, firmly in mind. (California District Attorneys Association, 1980; Galvin, Garner, Kluger, and Toscano in Hammett, 1987).

Evidence on fire cause and origin can be complicated and confusing, Moreover, these complexities may be exploited by defense counsel to induce doubt in the minds of jurors regarding the incendiary origin of the fire. Thus, prosecutors should prepare their cause-and-origin witnesses:

- "to make their court presentations as clear and intelligible as possible" (Hammett, 1984). The prosecutor should ensure that the expert witness can speak effectively in layperson's terms--that is, avoiding technical jargon (California District Attorneys Association, 1980).

- to be an **entertaining** witness who can tell the story effectively and hold the jury's interest (Burnette and Smith, 1980).

The prosecutor should also obtain from all expert witnesses full resumes of their qualifications well in advance of trial. These should include:

- formal degrees;

- other education and training programs attended;

- publications;

- job experience, and

- court qualifications as an expert witness (Campbell in Hammett, 1987).

This information will assist prosecutors in qualifying experts at trial. Prosecutors should pay particular attention to the witness' practical experience in fire investigation, which is often more extensive and important than the individual's formal educational background.

Trial Tactics

The following are general recommendations for approaching an arson trial (California District Attorneys Association, 1980):

- **develop a theme** or short phrase for repeated use in describing the case;

- **identify a victim** with whom the jury will sympathize: an injured firefighter; elderly tenant; child's pet; but not an insurance company;

- **get excited** about the case;

- **get organized--e.g.,** break the presentation of the case down into major "acts"; prepare detailed outlines of the testimony of each witness cross-referenced to relevant documents;

- **stress the seriousness of the crime** of arson throughout the court presentation. As noted earlier, many potential jurors and even judges consider arson to be primarily a property crime in which the main victim is the insurance company;

25

- assemble and use at trial a list of basic facts and statistics on arson that clearly demonstrate its **devastating human, economic, and social impacts;:**

  --deaths and injuries to civilians and firefighters;

  --lost tax revenues;

  --urban decay and deterioration;

  --lost housing stock;

  --lost jobs and earnings;

  --increased insurance premiums for everyone.

- Make the point should be made that **arson ultimately effects everyone in the community;**

- Plan ways to **"transport" the jury to the scene** of the fire. One experienced arson prosecutor stresses the importance of presenting evidence in all sensory dimensions--sight, smell, sound and touch--that evokes the fire scene as persuasively and graphically as possible;

- Use to best advantage a **powerful hero figure-the firefighter** (Calvin in Hammett, 1987).

Selecting the Jury

Prosecutors should give careful consideration to selecting types of jurors most likely to be responsive to the state's arson case. Obviously, jury selection must be tailored to each particular case. Ironclad principles cannot be enunciated. However, prosecutors may wish to consider the following list of potentially desirable juror types in arson cases (Galvin in Hammett, 1987; Burnette and Smith, 1980):

- fire victims or relatives/friends of fire victims;

- residents of the neighborhood where the fire occurred or of neighborhoods with serious fire problems;

- people who have read about the arson problem;

- applied scientists (they will be likely to identify with the cause-and-origin investigator);

- firefighters or relatives of firefighters (they are part of a "brotherhood that sticks together");

- insurance company employees;

26

- people with accounting or banking backgrounds, particularly in complex arson-for-profit cases;

- well-educated people (because arson evidence is often complex);

- political liberals, particularly in landlord arson-for-profit cases (because they are extremely hostile to slumlords).

- elderly women living alone, particularly in spite-and-revenge cases.

Some prosecutors have expressed concern that several of these juror types--in particular, victims and their relatives/friends and firefighters and their relatives/friends--are likely to challenged by the defense. Moreover, there is some disagreement about the desirability of selecting scientists or persons with a scientific bent to arson juries. One arson prosecution guide suggests that such individuals may tend to second-guess and be suspicious of the government's cause-and-origin witness. This guide also recommends that any prospective jurors who express reservations about cases built on circumstantial evidence and cases built on co-conspirator testimony should be peremptorily challenged (California District Attorneys Association, 1980).

Making Broader Use of Jury Selection

If it is permitted in their jurisdiction, prosecutors should also take a broader view of the jury selection process. It should be considered not only as a means to select jury members but also as an opportunity to educate the jurors about arson and cases relying on circumstantial evidence. The following points should be made:

- arson is not just another category of fire, but rather a violent crime;

- circumstantial cases are often stronger than cases resting on direct testimonial evidence; and

- (in cases relying to any degree on co-conspirator testimony) anyone is capable of lying or telling the truth--whether the person is a police officer or a convicted drug dealer--and that decisions regarding the veracity of a particular witness should not be based on blanket predispositions.

Opening Statement

According to an experienced arson prosecutor, the opening statement should never be waived. This is the opportunity to have "first crack at the jury." The statement should (Galvin in Hammett, 1987; California District Attorneys Association, 1980; Burnette and Smith, 1980):

- summarize in "understandable terms" the **key testimony** that will be offered by firefighters, cause-and-origin investigators, laboratory chemists, insurance officials and other witnesses;

- tell **"a good story"** that keeps the jury's attention;

- **never confuse the jury.** This would be a serious blow to the government's case at its very outset;

- **avoid legal and law enforcement jargon;**

- begin **"personalizing"** the victims;

- **"get the jury's attention"** by "making them see fire and smell smoke." Show the jury photographs of the fire and its victims and samples of fire debris (if exhibits can be marked in advance and the judge will allow their use during the opening statement), exhibits can be marked in advance and the judge will allow their use during the opening statement,

- **appeal to the fear of fire** which is so prevalent in the United States.

In sum, the opening statement is a time to give the story some drama. On the other hand, the strength of the case should not be overstated in the opening, as the jurors will quickly recognize the discrepancy later on. This can call the credibility of the prosecutor's entire case into question (California District Attorneys Association, 1980).

Order of Witnesses and Evidence Presentation

There is some disagreement among experienced arson prosecutors as to who should typically be the first witness and the decision must be based on the situation presented by each case. Again, it is not possible to enunciate ironclad principles. Some argue for ordering the prosecution witnesses according to the elements of arson--that is, first witnesses establishing that a fire occurred, followed by witnesses establishing that the fire was incendiary, and finally witnesses establishing the defendant's connection with the fire (California District Attorneys Association, 1980). Compatible with this theory is the advice offered by one prosecutor: "firefighters first." Firefighters are the "last American heroes" and their appearance on the stand in uniform, using their terminology to carry the jury "from the firehouse to the inferno" can "make your case." According to this view, firefighter witness can effectively set the stage for the cause-and-origin expert, by providing a general description of the fire scene and the first observed indications of incendiary origin (Galvin in Hammett, 1987).

On the other hand, California arson prosecutors guide recommends putting the victim on the stand first. The most lasting impression of the case should be presented first, according to this argument. The most telling question of this witness should be the last one asked: "how many people were asleep in this building when the fire started?" (California District Attorneys Association, 1980).

In other cases, it may be desirable to call neither a firefighter nor the victim first. For example, if the defendant told a particular witness of his intention to burn his building, it might be advisable to call this witness first,

There are several other considerations which affect the ordering of witnesses and introduction of evidence. For example, if a confession is to be introduced, the corpus delecti of the crime must first be proven. However, the Florida prosecution guide offers some good practical advice on introducing evidence that simultaneously establishes corpus delecti and links the defendant with the fire. This will almost inevitably be necessary because the arson investigator will offer testimony not only on incendiary origin, but also on the agency of the defendant--such as the defendant's purchase of a flammable liquid (Burnette and Smith, 1980, pp 4B-8 to 4B-12).

Another issue concerns co-conspirator witnesses. Because of the perception that their testimony may be tainted by ulterior motives, co-conspirators should be "sandwiched" by more credible witnesses (Galvin in Hammett, 1987). Finally, the California prosecutors guide recommends that the last witness be a civilian to 'rehumanize" the case after "impersonality" of government witnesses (California District Attorneys Association, 1980).

Tactics for Particular Witness Types

Firefighters. Experienced prosecutors emphasize that firefighters should appear in full uniform at trial and sentencing. This takes full advantage of their "hero image" and helps to distinguish them from police officers who may be distrusted by jurors. Firefighter witnesses should:

- Describe all of their activities at the fire scene in their own words. They may become confused or flustered if pressured to use other, more technical or scientific terms; and

- Explain any obscure or unclear terminology.

Firefighters can sometimes offer very effective testimony on the cause and origin of the fire, based simply on their observations of the color of the smoke and the intensity of the fire and the condition of doors and windows. Key observations to elicit from firefighter witnesses include (Burnette and Smith, 1980, 5F-1 to 5F-3):

- wind conditions;
- color of smoke;
- rapidity of spread;
- initial observations of people at the scene;
- means of entry;
- number of areas of fire;
- intensity of heat;
- unusual odors);
- condition and contents of the building (furniture, closets, holes in floor, sprinkler system operation, fire doors);
- possible ignition devices;
- trailers; and
- condition of appliances and electrical system.

Firefighter testimony regarding the use of mouth-to-mouth resuscitation to revive children overcome by smoke and "covering exposures" (i.e., preventing the fire from spreading to neighboring properties) may also be very helpful to the prosecution, particularly if these buildings are occupied by elderly or otherwise vulnerable persons. Such testimony may help capture the sympathy of the jury and establish endangerment to others and/or other property where this is an element required by the criminal charge in the case (Galvin in Hammett, 1987), but it may be inadmissible as irrelevant and prejudicial in some cases.

Cause-and-Origin Investigators. The first and most important step with cause-and-origin witnesses is to qualify them as experts. We have already discussed steps to prepare for this process, such as submission of full resumes to the prosecutor. The California and Florida arson prosecutors guides provide useful lists of questions to use in qualifying cause-and-origin experts (California District Attorneys Association, 1980, pp X-16 to X-23; Burnette and Smith, 1980, pp. 5G-1 to 5G-2). These cover:

- formal training and education;
- extent of actual investigative experience (e.g. number of fires investigated);
- previous expert testimony;
- knowledge of:

--specific fire causes,

--accelerants,

--burn patterns, and

--technical terminology of fire scene examination.

Once qualified, witnesses testifying on the cause and origin of the fire should present an logical and intelligible (to the layperson) explanation of fire behavior and of the steps involved in fire scene examination. Ideally all evidence relied on in the cause-and-origin determination should have been introduced before the cause-and-origin expert is called. His testimony should then move carefully and logically toward the cause determination. The California guide offers examples of question sequences covering the process of cause-and-origin determination, elimination of accidental causes, and opinions as to fire cause. These should be carefully prepared and rehearsed in advance (California District Attorneys Association, pp X-23 to X-24).

The cause-and-origin witness should use photographs, videotapes, diagrams, and models if such exhibits unequivocally support his opinions. The prosecutor's use of such exhibits should be carefully planned and orchestrated so that the government effectively controls the sequence in which evidence goes to the jury.

It may be useful to the prosecutor to have an "advisory witness" available for consultation during the entire trial, or at least those portions relating to the cause and origin of the fire. Depending on state law or court rules, this may be the same person who testifies for the state on cause and origin or another expert in the field. This advisory witness may be able to assist the prosecutor both in presenting the state's case to best advantage and in rebutting the defense's theory of cause and origin (Hammett, 1984).

Victims. The most important role of a victim witness is to humanize the effects of the fire. This should be done through testimony on their discovery and reporting of the fire and their description of the fire. The latter might help buttress the state's case regarding incendiary origin (Calvin in Hammett, 1987).

Victims might also be able to offer valuable testimony on the activities of the defendant in the period before the fire. If the victim observed the defendant removing valuable items or fixtures before the fire, this can help to link the defendant to the fire (Calvin in Hammett, 1987).

Co-Conspirators. Co-conspirator witnesses may be indispensable to the state's case but they also pose serious credibility problems for the prosecution. In general, they are probably the least believable government witnesses. The following guidelines may assist the prosecutor in maximizing the value--and avoiding the pitfalls--of co-conspirator testimony (Calvin in Hammett, 1987):

- Obtain from the co-conspirator a written agreement to testify, but one which commits to no specific sentence recommendation. If the witness' credibility is questioned, this agreement can be produced in court.

- The co-conspirator must clearly admit that "he's a bad guy" whose only hope for a break depends on being "100 percent truthful." Claims of having "gotten religion" are usually not believable.

- The co-conspirator should testify only to facts that can be independently corroborated by other witnesses.

- The prosecutor should not be friendly to the co-conspirator witness, but rather should treat this individual with professional detachment.

Laboratory Chemists. A key issue for prosecutors is the advisability of presenting evidence of laboratory analysis with negative or unclear results. Some prosecutors advise investigators against sending samples for laboratory analysis unless they are quite sure the samples will yield positive results. According to this view, it is easier to defend a failure to send samples for analysis than to defend a negative or unclear test result (Kluger in Hammett, 1987). Other prosecutors argue that it is safer to present the negative or unclear results and have the laboratory expert explain the possible reasons for these results than to allow the defense to make the point that the state did not even bother to have samples analyzed. The fact that samples were taken and analyzed helps to show the jury that a through and professional scene examination was conducted (Calvin in Hammett, 1987).

If the prosecutor decides to put a chemist on the stand, this witness must be qualified as an expert. The Florida arson prosecution guide provides a checklist to assist in qualifying chemists (Burnette and Smith, 1980, pp 5G-3 to 5G-4). This covers:

- job position;
- specialization;
- certifications;
- length of experience;
- education and degrees;

- specific related experience;
- publications; and
- previous expert testimony.

In his or her testimony, the laboratory chemist should:

- **explain the tests and procedures** used to analyze the fire debris in terms understandable to the layperson;

- **make clear what conclusions can and cannot be drawn** from the tests. Most laboratory chemists will be unwilling to testify that a sample definitely contained gasoline or any other particular flammable liquid. Rather, they will testify that the sample contained a liquid "consistent with" or "similar to" that flammable liquid. The reasons for this apparent hesitancy should be clearly explained to the jury (Hammett, 1987).

Insurance Witnesses. The most common insurance witnesses are the insurance agent and the private cause-and-origin investigator hired by the company. If the case is alleged to involve an insurance fraud, the agent may testify to the defendant's false statements on the insurance application and attempts to overinsure the property. The insurance company's investigator can be used to corroborate the public investigator's theory of the fire's cause and origin, assuming that the two coincide. However, when the reports of the two investigators are inconsistent, the prosecutor may face serious problems in convincingly presenting the state's case (Hammett, 1987).

Physical Evidence from the Fire

Physical evidence may be the key to the prosecutor's effort to have the jury experience the fire as closely as possible, understand the cause, and link the defendant with the crime. Possible items include:

- samples of fire debris;
- ignition devices;
- fingerprints; and
- defendant's clothing.

It may be effective to pass a sample of debris from the fire that retains a strong odor of gasoline or other flammable liquid; if such a sample is available. Samples of fire debris should be passed in closed metal paint cans rather than clear plastic bags. The cans create more of a sense of mystery and they must be opened to view the contents.

This gives the jurors the opportunity to smell the fire and the vapor from the flammable liquid. A clear plastic bag need not be opened to view the contents and thus the opportunity to have the jury smell the fire and flammable vapors may be lost (Galvin in Hammett, 1987).

The ignition device, if found, is a critical piece of physical evidence. The jury should be given the opportunity to look at the device and, if permitted by the court, the prosecutor should conduct a practical demonstrations of its workings. This will help the jury to understand the prosecution's theory of incendiary origin (Calvin in Hammett, 1987). An even more advantageous approach is to take the jury to the actual fire scene and conduct test recreations of the fire.

Proper chain of custody must be demonstrated for all physical evidence introduced at trial. "Predicate" problems may also arise in the case of evidence found and collected at the fire scene on a visit subsequent to the original scene examination. The defense might argue that the scene was not necessarily in the same condition at the time the evidence was collected as it was at the time of the fire and initial search. These issues again underline the need to maintain security of the fire scene for as long as possible (Texas District and County Attorneys Association, 1980).

In addition, locked doors and windows can be good evidence against forced entry and in favor of fraud arson by the property owner (California District Attorneys Association, 1980). In order to meet any defenses alleging forced entry by an arsonist, the prosecutor should know exactly which doors and windows were broken and entered by firefighters in a locked building. This should be supported by testimony from the firefighters involved.

Tape Recordings, Photographs, Videotapes, Diagrams and Architectural Models

Although there may be restrictions on their use in some jurisdictions, tape recordings can be the prosecutor's most important real evidence. They can prove that the defendant lied and/or made inconsistent statements. Recordings of the fire being reported, particularly if they reveal panic or fear on the part of victims, may also be very effective in convincing the jury of the seriousness of the crime (Galvin in Hammett, 1987).

Photographs and videotapes may be very useful in presenting the state's case on the incendiary origin of the fire, However, they also may serve as important tools in appealing to the emotions of jurors. Some prosecutors believe that black-and-white photographs are best for documenting the critical aspects of the fire's cause and origin, while others argue that color photographs are preferable when trailers must be documented; color is more sensitive to shadings on a surface than is black-and-white. It is generally agreed that color photographs are more effective in showing overall views of the fire and of the firefighters working to contain it. Color photographs are also best for "poignant scenes" involving victims and firefighters (Galvin in Hammett, 1987). As noted earlier, however, some jurisdictions may exclude wide-angle or telephoto shots unless specific reasons for their use can be presented to the court.

Investigators and prosecutors should also offer diagrams and models of the fire scene, particularly if government's theory of incendiary origin is complex. These models and diagrams may be useful both in explaining the cause and origin of the fire and in helping the jury to visualize the state's case regarding the conduct of the defendant. Architectural models can be more effective than diagrams in conveying the physical environment of the fire scene. Moreover, the fact that the prosecution took the trouble to build a model helps to convince the jury of the importance of the case. Models may often be obtained at fairly reasonable cost from small architectural firms (Galvin in Hammett, 1987). Of course, it is very important that all diagrams and models be absolutely accurate. There should be no opportunity for the defendant or a defense witness to expose inaccuracies in these exhibits. (California District Attorneys Association, 1980).

Meeting Defenses

Defense Cause-and-Origin Witness. In the federal system and many states, experts are quite easily qualified. However, in some jurisdictions, the best way to meeting the defense's theory of cause and origin may be to attack the qualifications of their witness. Prosecutors should ask pointed questions of the witness, such as the following (Calvin in Hammett, 1987; California District Attorneys Association, 1980--see pp X-27 to X-31 for an extensive list of questions for cross-examination):

- What is your training in cause-and-origin determination?

- How many complete cause-and-origin investigations have you done?

- Have you ever given testimony on the entire cause-and-origin determination of a fire rather than simply on the elimination of one accidental cause?

- Have you ever testified as to cause-and-origin without personally examining the fire scene?

Cross-examination based on such questions may disqualify (or at least call into question) many defense cause-and-origin witnesses (Galvin in Hammett, 1987).

Even if the defense witness does qualify as an expert, his testimony can often be challenged by the following strategies:

- Testing their knowledge of the details of fire investigation and laboratory analysis. If they make errors, their credibility in the eyes of the jury may be undermined.

- Pinning down if and when they personally examined the scene (defense cause-and-origin experts usually do not see the fire scene until much later than the state's investigators, if at all). If there was no immediate examination of the scene, the witness will probably have to admit under questioning that a proper cause-and-origin determination cannot be done based only on photographs, which may be misleading, or on a later examination, after the condition of the scene may have changed.

- Determining whether they personally interviewed any witnesses-- firefighters, police officers, victims, others-- in this case. (Galvin in Hammett, 1987; California District Attorneys Association, 1980).

As noted earlier, an "advisory witness" who is a qualified cause-and-origin expert may be able to assist the prosecutor in attacking the testimony of the defense's witness. This advisory witness may be the state's cause-and-origin witness or another expert familiar with the case, depending on state law or court rule regarding the presence of witnesses in the courtroom during the testimony of others (Hammett, 1987).

Finally, some prosecutors keep "libraries" of transcripts of the testimony of defense cause-and-origin witnesses, The number of these witnesses is relatively small, and they tend to appear in many cases and often to rely repeatedly on the same theories of fire cause. Having transcripts of previous testimony available helps the prosecutor to prepare cross-examination strategies and to combat defense attacks on prosecution witnesses.

Alibi Witnesses. Tactics for meeting alibi defenses in arson cases should be the same as employed in other criminal cases--the key should always be repeated questioning on details so as to demonstrate inconsistencies in various versions of the story given by the witness. Repeated tape recorded interviews prior to trial are critical to this strategy. Prosecutors may be able to show that the alibi witness is unshakable on areas that have been rehearsed with defense counsel but may dramatically lose accuracy and consistency when the questioning ventures into unrehearsed parts of the story (Galvin in Hammett, 1987). A potentially effective strategy is to establish that the defendant developed an alibi before he or she knew the exact dates and times that had to be covered (California District Attorneys Association, 1980). Prosecutors should also stress any relationship between the witness and the defendant and closely question any delays in the witness' coming forward with his or her story (Galvin in Harnmett, 1987).

Intoxication. A common defense in arson cases is that the defendant accidentally started the fire while intoxicated. Prosecutors can often effectively meet this defense by demonstrating that an intoxicated person could not have done the things necessary to set the fire or, at least, could not have done so without suffering a severe injury in the process (Galvin in Hammett, 1987).

Pathological Firesetting and Mental Responsibility Defenses. In many arson cases, defenses based on mental condition will be advanced. The California prosecution guide provides a detailed discussion of the legal and psychological concepts involved in such defenses and suggests strategies for effectively meeting them (California District Attorneys Association, 1980). This includes

- discussion of pyromania as mental disease;

- mental impairment (which might include intoxication, or unconsciousness);

- mental incapacity to form general criminal intent (actus reus);

- substantial mental impairment; and

- insanity defenses.

Prosecutors can successfully meet defenses based on mental impairment or mental incapacity by demonstrating that the defendant was able to perform "customary and usual social functions." Relative to setting the fire, the prosecutor should try to show that the defendant

- understood the physical nature and meaning of the act and its consequences--for example;

- planned the firesetting act in advance;

- pursued a plan in a goal-directed way, used suitable materials;

- selected a place to set the fire that was not immediately visible; and/or

- took steps to avoid apprehension (California District Attorneys Association, 1980).

Many of the same showings can be used to meet an insanity defense. For example, evidence that the defendant consciously sought to avoid apprehension implies an appreciation that the act was wrong and illegal and can be used to rebut an insanity defense, even if the defendant cannot be shown to have specifically addressed the illegality or wrongness of the act (California District Attorneys Association, 1980).

Closing Argument

In the closing argument, the prosecutor has another important opportunity for the prosecutor to exercise persuasiveness. It should be used to

- summarize and emphasize the evidence of incendiary origin (particularly the sensory physical evidence);

- catalogue the defendant's lies or inconsistent statements; and

- reemphasize that absolute proof is not necessary to convict, only proof beyond a reasonable doubt (Calvin in Hammett, 1987).

All experienced prosecutors recognize the importance of finishing the case in a strong and persuasive manner.

JUVENILE CASES

Juvenile arson is a serious problem. A substantial number of arson fires are probably set by persons classified under the law as juveniles. It is difficult to discuss the handling of juvenile arson cases from a national perspective because juvenile justice systems differ so widely across jurisdictions. However, most systems recognize four gradations in the processing of juvenile offenders (Hammett, 1984):

- "lecture and release" by police;

- informal diversion" through counseling, education, or other programs;

- prosecution in juvenile system--outcomes include counseling, supervised probation, and confinement in a youth facility; and

- bindover for prosecution in adult court--this is usually reserved for serious offenses.

Most prosecutors see juvenile arson as a problem requiring different responses from those applicable to adult arson. Indeed, in most jurisdictions, juvenile cases are assigned to an entirely different prosecutorial unit. Prosecutors involved in juvenile cases should be sensitive to youths' normal lack of sophistication and frequent lack of understanding of the implications of their actions, At the same time, prosecutors must be vigilant for possible relationships between juvenile and adult arson cases. In particular, if a "torch" is being prosecuted in juvenile court and the individual who hired the torch is being prosecuted in adult court, it is important that the prosecutions be closely coordinated so that valuable evidence is not lost to either effort (Hammett, 1984).

APPENDIX A:

GUIDE TO ISSUES AND SOURCES

Figure A.1

ARSON PROSECUTION : MATRIX OF ISSUES AND SOURCES

Sources of Further Information (See References for Full Citations)

Issues	Hammett, 1984	McClees, 1983	Burnette and Smith, 1980	Calif. District Attorneys Association. 1980	Maryland State Attorneys Coord, 1981	Texas District and County Attorneys Assoc. 1980	ATF/NIJ Arson for Prosecutors, 1985
Role of the Prosecutor in Arson Investigation	135-139.247	4-89 to 4-92	1 B-1	1-9.1-10			
Examining the fire scene		3-37 to 3-45	1 B-2	x-2		Ch. VII	
Legal issues		3-37 to 3-45	Sec. 1C		73-78	Ch. VI, esp. pp. 53-59	
Cause-and-Origin determination	101-106		1B-2 to 1B-6, 1B-8 5A-1 to 5A-5	35-44,95,117-125			Toscano, Arson Investigators Checklist; Campbell, Cause-and-Origin Determination
Documenting the scene			1B-9 to 1B-11			Ch. VIII	Toscano, Arson Investigators Checklist
Collecting samples of fire debris		3-46 to 3-59	1B-6 to 1B-8	XIII-1 to XIII-2	44-49	Ch. VIII	Campbell Cause-and-Origin Determination; Garner, Gas Chromatography
Developing the Criminal Case							
Identifying and interviewing witnesses			5A-6				Toscano, Arson Investigators Checklist; Paneton, Sources of Interview Information
Developing evidence of motive	106-109			Ch. V	125-145	Ch. IX, esp. pp. 89-105, 114-115	Toscano, Arson investigators Checklist; Strawbridge, How to Handle an Arson and Fraud Case
Developing investigation reports	246-247	3-102 to 3-109, 4-93 to 4-95					
Timing arrests	153-154						

Sources of Further Information (See References for Full Citations)

Issues	Hammett, 1984	McClees, 1983	Burnette and Smith, 1980	Calif. District Attorneys Association, 1980	Maryland State Attorneys Coot-d, 1981	Texas District and County Attorneys Assoc, 1980	ATF/NIJ Arson for Prosecutors, 1985
Arson Prosecution							
Structuring arson prosecution	237-239			I-6 to I-8			
Screening cases for prosecution	145-199	4-96 to 4-106		Ch. VII			
Determining the mode of charging			Sec. 3A	Ch. VIII-IX			
Applicable statutes	28-38	Section 4.1	Sec. 2A,2B	Ch. IV			
Preparing witnesses	253		3C-2	X-2, XIII-6	66-70	Ch. X	
tactics:				Ch. X	Ch. VII	Ch. X	
Selecting juries			4A-1	X-13			Galvin, Trial Techniques, PP. 1-2
Opening statement			48-1	X-13	99- 103		Galvin, Trial Techniques, p.2
Order of witnesses			4B-5 to 4B-6, 4B-8 to 4B-12	X-14 to X-15			Galvin, Trial Techniques, p.5
Tactics for witness types:							
Firefighters			5F-1 to 5F-3				Galvin, Trial Techniques, pp. 2-3
Cause-and-origin investigators	253		3C-2, 4B-7 to 4B-9, 5G-1 to 5G-2	X-2, X-14	103-112		Galvin, Trial Techniques, pp. 3-4
Victims							Galvin, Trial Techniques, p.4
Co-conspirators							Galvin, Trial Techniques, pp. 5-6
Laboratory chemists			5G-3 to 5G-4	XIII-6			Galvin, Trial Techniques, p.6

Sources of Further Information (See References for Full Citations)

Issues	Hammett, 1984	McClees, 1983	Burnette and Smith, 1980	Calif. District Attorneys Association, 1980	Maryland State Attorneys Cord-d, 1981	Texas District and County Attorneys Assoc. 1980	ATF/NIJ Arson for Prosecutors, 1985
Insurance witnesses						119-120	Galvin Insurance Witnesses
Physical evidence from fire				VIII-8, X-15 to X-16			Galvin, Real, Demonstrative, and Documentary Evidence
Recordings, photographs, videotapes, diagrams and models				X-16			Galvin, Real, Demonstrative, and Documentary Evidence
Meeting defenses							
Defense cause-and-origin witness				X-25 to X-31			Galvin. Trial Techniques, pp. 7-8
Alibi witnesses				X-34 to x-35			Galvin, Trial Techniques pp. 9-10
Intoxication				XII I-37 to XII-38			Galvin, Trial Techniques, p.8
Mental responsibility defenses			4B-14 to 4B-17	Ch. XII	112-115		
Closing argument				Ch. VI			Galvin, Trial Techniques, p.11
Juvenile cases	App. B						

REFERENCES

The following reports, sponsored by the National Institute of Justice, U.S. Department of Justice, are available from the National Criminal Justice Reference Service, P.O. Box 6000, Rockville, Maryland 20850.

Abt Associates Inc. (1980). Program Models: Arson Prevention and Control.

Hammett. Theodore M. (1987). Cracking Down on Arson: Issues for Investigators and Prosecutors. This document summarizes presentations by the following speakers at the NIJ/ATF Arson Prosecution Conference, Federal Law Enforcement Training Center, Glynco, Georgia, November 1985:

- Bruce Bogart, American Insurance Services Group

- David Campbell, North Carolina State Bureau of Investigation, Arson Div.

- Mary Galvin, Assistant State's Attorney, Judicial District of New Haven, Connecticut

- Dr. Daniel Garner, Chief, Forensic Laboratory, ATF National Laboratory Center

- Michael Martin, ATF Regional Counsel's Office, Atlanta

- Barry Kluger, Assistant District Attorney, Bronx County, N.Y.

- John P. Panneton, United States Attorney's Office, Sacramento, Calif.

- David R. Strawbridge, Esq., Cozen and O'Connor, Philadelphia

- Joseph P. Toscano, Inspector, New Haven (Connecticut) State's Attorney's Office

Hammett, Theodore M. and Kimberly A. Wylie (1987a). Issues and Practices: Toward Comprehensive Arson Control Programs.

Hammett, Theodore M. (1984). Arson Investigation and Prosecution: A Study of Four Major American Cities.

Ku, Richard, Theodore M. Hammett, et al., 1980. Arson Control: A Synthesis of Issues and Strategies Based on the Arson Control Assistance Program.

McClees, Hugh C., Andrew J. Decker, and Daniel J. Carpenter (1983). Managing Arson Control Systems: Executive Summary (Daniel Ford, ed.). (The full report--Managing Arson Control Systems: Arson Detection (vol. 2), Arson Investigation (vol. 3,) Arson Prosecution (vol.4), and Appendices (vol. 5)--is available on loan.)

New York City Arson Strike Force (1983). <u>A Study of government Subsidized Housing Rehabilitation Programs and Arson: Analysis Programs Administered in New York City, 1978-1981.</u>

Tauber, Stephen J. (1978). <u>Arson Control: A Review of the State of the Art with Emphasis on Research Topics.</u>

Webster, Stephen H., and Kenneth E. Matthews, Jr. (1979). <u>A Survey of Arson and Arson Response Capabilities in Selected Jurisdictions,</u>

The following state arson prosecution guides are also available through the National Criminal Justice Reference Service:

Burnette, Jr., Guy E. and Lawrence W. Smith (1980). <u>Florida Arson Prosecution: A Trial Manual for Florida Prosecutors.</u>

California District Attorneys Association (1980). <u>Arson Prosecution.</u>

Maryland State's Attorneys Coordinator (1981). Maryland State's Attorneys Arson Investigation and Prosecution Manual.

Texas District and County Attorneys Association (1980). Arson Prosecution.

Other Materials

American Bar Association, Young Lawyers Division, Arson Committee. The Comrnittee publishes the Arson Reporter (current caselaw) and has issued an Arson Legislative Reference Manual and an Arson Legal Resource Directory. For information, contact the Committee at 750 North Lakeshore Drive, Chicago, Illinois 60611.

Factory Mutual System, <u>A Pocket Guide to Arson Investigation</u> (2d edition: Norwood, MA, 1979). Available from Factory Mutual Engineering and Research, 1151 Boston-Providence Turnpike, Norwood, Massachusetts 02062.

Gaensslen, R.E. and Henry C. Lee, <u>Physical Evidence and Forensic Science,</u> (2nd ed.: American Insurance Association, 1986).

Gelband, Barbara (1980). "The Prosecutors Role in Fire Investigation." Presented at National College of District Attorneys Seminar on Arson Investigation and Prosecution

Pisani, Angelo (1982). "Identifying Arson Motives," <u>Fire and Arson Investigator</u> 21(June 1982), pp 18-24.

U.S. Department of the Treasury, Bureau of Alcohol, Tobacco and Firearms, <u>Arson Case Briefs</u> (Washington, November 1985). Available from BATF Explosives Division, Arson Enforcement Branch, 1200 Pennsylvania Ave., N.W., Washington, D.C. 20226).

U.S. Department of the Treasury, Bureau of Alcohol, Tobacco and Firearms and

U.S. Department of Justice, National Institute of Justice, Arson for Prosecutors. Materials assembled for the NIJ/ATF Arson Prosecution Conference, Glynco, Georgia, November 1985.

 *U S Government Printing Office: 1994 - 520-381/81079

www.ingramcontent.com/pod-product-compliance
Lightning Source LLC
Chambersburg PA
CBHW081229170526
45165CB00009B/3005